ELECTRICAL STEELS form
a vital part of modern electrotechn[
are used to secure static and rotatir
of economic size and efficiency.

Prof. PHILIP BECKLEY has spent 45 years at
the industrial forefront of Electrical Steel
development as well as lecturing widely in
University and conference venues round the
world. He brings these insights onto your
desk.

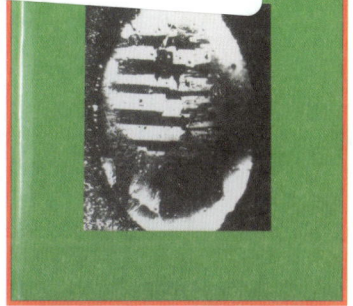

This book mediates between the worlds of
metallurgical steel processing and the electrical applications of core steels.
Read by both, it promotes constructive communication and facilitates
co-operative projects.

A comprehensive chapter on measurement and test complements those on
Basics, Coatings, Heat Treatments, Motors, Generators, Domain activity,
Process Routes and much more. A comprehensive set of steel
characteristic curves is included as well as an extensive glossary.

This book runs to some 500 A4 pages and is profusely illustrated by
hundreds of diagrams, tables and colour photographs. This is an enduring
guide and reference work full of interest and insider views.

It is presented in hardback format with dustjacket.

£75 + post and packing. ISBN 0-9540039-0-X

This is a limited edition.

To enquire about signed copies Tel 44 (0)1633-853906

The Effective Engineer

This book is dedicated to Stephen Best.

"O wad some Pow'r the giftie gie us to see
oursels as others see us!
It wad frae mony a blunder free us,
And foolish notion." *Robert Burns*

The Effective Engineer

A career aid for Graduate Engineers.

by Prof. Philip Beckley D.Sc. FIET. C.Eng.

Bettws Books 2007

First Edition 2007

Copyright Philip Beckley © 2006

All rights reserved. No part of this publication may be reproduced, stored in a retrieval system, or transmitted at any time by any means electronic, mechanical, photocopying, recording or otherwise, without the prior permission of the publisher.

ISBN 978-0-9556552-0-3

The contents of this book are produced for the interest and education of readers and neither the author nor those providing input to the work can be responsible for loss damage or any other detriment, which may arise from use of the material within this book, which has been provided in good faith.

Published by Philip Beckley, Bettws, Newport, South Wales, NP20 7AD.
Printed in (2007)
by HSW Print Tonypandy, Rhondda CF40 2XX

Acknowledgements

I am indebted to MARY BECKLEY for her helpful work on the computer which made this book possible.

My thanks go to Nicolette Pace who procured for me some original concept illustrations via her colleague Hanneke Kools in Holland.

I am grateful to Dr. S. Zurek for generous help with proof reading.

Preface

Student Engineers have committed themselves to a stringent course of study aimed at providing a toolkit of technological techniques, which will, when supplemented by updates as years pass support their activities in employment for a lifetime.

Undergraduates are well aware that this means a lengthy commitment to hard work and learning, and they are mentally prepared to undertake it. They will have 'swotted' for A-levels and are ready to apply similar effort to undergraduate engineering. They realise that evening enjoyment and social activity must be tailored so that sufficient effort goes into lectures, assignments and reports.

After 3+ years of effort students expect to get a degree with a respectable grade and to find that this is acceptable to employers.

In the world of sport it is well known that simply running the course is seldom enough and that for stardom a study must be made of what is needed to excel.

To be a sought-after engineer also requires sustained training in relevant techniques.

This little book tries to illustrate the paths leading to excellence.

Most students have never had any training in Life Skills, and if asked how much has been placed before them are likely to say "none that I noticed".

The Effective Engineer

Foreword

Graduates have a lot on their minds these days. The euphoria of graduation is soon overtaken by consideration of a debt mountain, where to live, what about getting a job and many more.

When at last a job is hooked the rest of life begins.

Are you ready for it? Have you spent time and effort during the undergraduate period getting ready, or are you like an office worker setting off to walk up Snowdon in winter with slack unready muscles clad in jeans and T-shirt?

Dear Graduate/Undergraduate please absorb from this text the armour needed to scale the heights in Industry. As a bonus find it enjoyable as well!

Contents

The Effective Engineer

Chapter 1
Examinations and Tests

● **A-Levels**

● **Bachelor Degrees**

● **Ph. D. exams**

● **Driving tests**

● **Hidden Tests**

1.1 A-Levels

The usual school examinations leads up to 'A' Level examinations. In times past these were all of the 'all in the head – write on the day' examinations.
Perhaps, the most demanding mode, see Fig 1.1.
Educationalists dispute long and loudly about whether this is better or worse than a modular and course-work approach.
Modular favours the bit-by-bit person rather than the gestalt operator and course-work should reveal steady application (see Fig1.2) but could involve desperate deadline meeting effort, or temptation to have uncle Gerald help and polish course-work a lot. However material which I learned 50 years ago still remains memory-accessible today and this facility astonishes students in this century. Is it useful or a sheer waste of memory space?

Midnight Oil Revision

Drip Feed of
Information

Tap to gush it all out on exam day

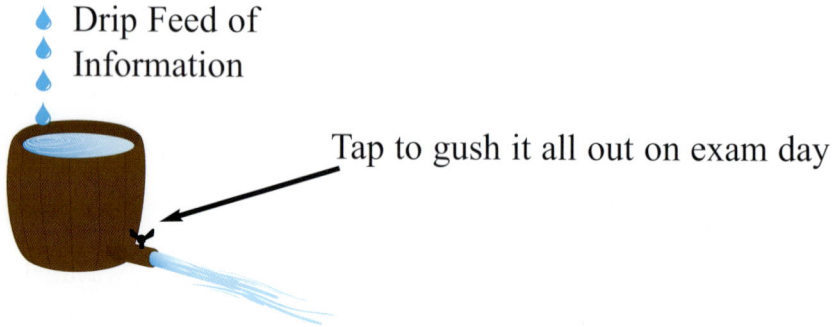

Fig 1.1 Traditional Exam System

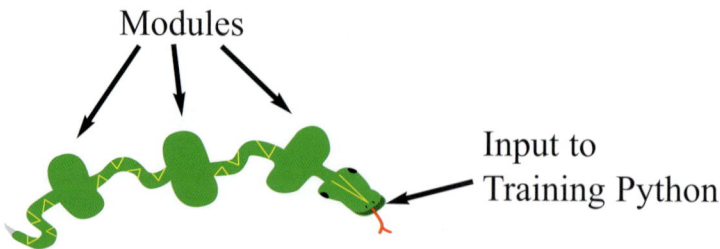

Modules

Input to
Training Python

Fig 1.2 Modular and Course Work Scheme
Parts digested slowly

1.2 **Degree examinations**

Again the slow integrator leading to one outpouring (Finals) is now working alongside the Modular mode.
Do the basics really sink in and stick?
I used to ask students "If you were knocked out in a car crash, when you came round in Casualty, if the Doctor said "What is the reactance of $1\mu F$ at 50 Hz?"
Would you say at once "About 3000Ω " and not need a calculator?

Some engineering courses require a project dissertation and this involves seeking data, organising it and writing presentably.
Can you do it?
Did you do it?
Are infinitives irreversibly split and do apostrophies baffle?
Does it matter?

When marking such work I am overjoyed at the sight of crisp correct English, though I may suspect uncle Gerald again.
A culture is growing up in which any information not instantly available on the 'internet' cannot be deemed to exist. If I show signs of using paper sources the Aged Anorak label is assured.

1.3 **Ph.D. Theses**

A whole book could be written on what these are supposed to be and do, and give guidance on how best to do it, - but I digress.

If you can coax an External Examiner to applaud your PhD thesis "well done" so be it. Especially non-native writers of English deserve praise. Some hints are given in ENVOI following Chapter 13.

1.4 **Driving Tests**

In this century (21st) the early acquisition of a driving licence is a great facility.

What is this sort of test like?

It's Multiple Choice questions, tick-the-box and a critically observed 'hands-on' evaluation.

Engineers are not often exposed to this type of shotgun evaluation in their job.

To pass you have to be knowledgeable, competent and cautious yet confident.

After a severe stroke I now drive an adapted car and had to undertake a stiff set of on-road driving and psychological tests. It's not easy to unlearn 50 years of accreted bad habits!

All these tests can help to characterise an Engineer and give a profile of capabilities. There also exist psychometric tests of many sorts of greater or lesser utility.

1.5 **Hidden Tests**

The purpose of this book is to acquaint the reader with the Hidden tests that can apply to an Engineer and show how to prepare for them.

Seldom spoken of, they operate throughout your life.

1.6 **The Car Park Test**

To explain this test let me set out a small scenario:

You are working in an Engineering Organisation, at any level and one day a Fire Drill is due. The Bell-Sounding ceremony is enacted and all troop into the Car Park, which is the assembly point.
The Department Manager and his Deputy pull rank and watch it all going on from the window (see Fig 1.3).
They start to talk about the New Product being developed that will need about 4 people to form a kernel team, and the Manager says, looking through the window "whom shall we pick to lead it and back them up?"
As the Fire Officer goes his rounds clipboard in hand the Deputy says "I think Bob Sims would be good, he's full of good ideas and always ready to help".

"Janet Brown would be a good worker - always on time and ready to do an extra hour if needed."

"What about Jock James?"-
"Can't rely on him, clever, but ask him to go into the jungle and shoot a lion and he may come back dragging a tiger over his shoulder. If I say "That's not a lion". He may reply "OK, but it's a really lovely tiger don't you think". When I want lions I want their leonine selves not stripy stand-ins.
They chat on looking over the Department team and end up saying "Well anyway don't put Dobbs or Grainger on it, they usually present late and don't get much done".

Fig 1.3a   The Car Park Test

The view you take of people depends on whose shoulder you look over –
see figure 1.3b

Fig 1.3b   The Car Park Test

By the time the Fire Drill is over preliminary choices have been made and those unspoken for in the Car Park could well figure in the next round of cuts that seem to come round in every business.

Can you? Do you prepare to do well in the Car Park test? Or is it something that happens which you know nothing about?
Personnel (human resource) officers would of course go white at the thought that it happened – but it does.

An early realisation of the likely existence of the Car Park test and how to be chosen in it is a vital career tool for the young student.

If it is true that employers recruit on qualification but fire on attitude we shall look carefully at what that implies for training.

1.7 The "Whom Shall We Send?" Conundrum

This is another 'informal' test usually conducted under pressure.
Here no new product is envisaged, but a customer problem has arisen.
Scenario:- A batch of inappropriate material has been sent to a customer and he is naturally displeased – "Whom can we send to sort it all out?"
Unfortunately the Department Manager is in hospital (sudden appendicitis) and his Deputy is in the middle of a two-week vacation in Florida.
"Whom can we send ?"
Dawkins? What about him?
We need someone who will chart the problem, get the facts and explain that we wish to get the problem sorted out, but will not make rash or expensive promises.
We need someone objective, likeable but firm.
Did you get training in such skills? Or was there no time due to battling with integration of complex variables?
The trouble with Dawkins though is that he could well "agree" with the customer and overegg their right to be fed up implying that our material is

basically poor.

No, we want firm balanced representation.

So whatever you do, don't send Bright, he could well snarl and make enemies we don't need.

Whom do you think gets sent?

Inevitably someone the acting manager feels he can trust, who will look after his company as well as being supportive of the customer.

How do you get to be the one to go?

It could be the first step on a road to becoming an important and trusted member of the Company.

In Chapter 2 we will look at the shape of Present-day Engineering Training and examine what we can do to give you a more complete toolkit of life skills

Fig 1.4 gives a view of the 'fixer' on his way to do battle.

Fig 1.4 Whom are they sending?

Questions

Before diving into Chapter 2, ponder these:-

a) Run over in your mind what sorts of test and examination you have experienced in life so far.

b) How well prepared were you for these?

c) Do you feel a 'natural' to be selected for things?

Chapter 2
Courses of preparation – The shape of Engineering Training

● **Courses of Preparation**

● **Undergraduates**

● **Postgraduates**

2.1 Outline. I see that it should look like Fig 2.1

We are all familiar with the Hardcore (left of Fig 2.1) pattern.
What about the right hand 'Cement' course content?
What might we expect to find in a life skills course aimed at fitting the student to become a top effective Engineer?
We shall look at an outline of course content and then discuss relevant areas in depth one-by-one.

Finding formal courses, which cover these areas is far from easy. Probably the interested reader may need to use the bibliography provided and make vigorous use of libraries and publications to fully cover the ground. The business faculty of most collages and universities have plenty of resources if the topics described are sought carefully.

2.2 Your present Situation

Unfortunately the cement side of training features all too faintly in an already bulging syllabus. The student/employee is thus urged to take charge of his/her own cement training and monitor its progress in a

process of self-evaluation.
A now time-hallowed technique for self assessment is the
SWOT Analysis (see Chapter 3 etc.)

HARDCORE	CEMENT
'A' level or equivalent Maths Physics Chemistry Metallurgy	SWOT Analysis (regular) Social interaction skills
	Hobbies
Basic Engineering – wide ranging	Community service
Engineering specialism	Volunteer work
I.T. skills	Professional Engineering Bodies
Continued Professional Development	Life skills

Figure 2.1

Figure 2.2 shows how the various inputs of a person's skills balance within their various situations during progress from A-level to Senior Rank.

A-Level is focussed on basics.

Undergraduate courses may, or may not, manage to include Life Skills.

Developing Engineers improve Life Skills either by hard experience or by judicious use of self-improvement material.

The development of taught skills during Engineering training and continued professional courses can look like this:-

'A' level course	
Engineering and Science Maths Other – various Life Skills	55 % 30 % 10 % 1-5 %
Undergraduate course	
Engineering Maths Others, inc IT Life Skills	55 % 25 % 15 % 1-5 %?
Graduate Engineer in employment post-grad courses	
Engineering- from job Other- from job Life Skills from job etc.	50 % 30 % 20 %
Senior Engineer with demanding job	
Cont. devel. Eng. courses various Languages Life Skills by experience	60 % 25 % 15 %

Fig 2.2

Questions

What training courses have you participated in since 'A' Levels?

Do you feel any specific lack of training?

Chapter 3
Self Evaluations

We can now give attention to a very important analytical process which can help with the choice and prioritisation of self training efforts as well as being a valuable tool applicable to many situations in employment.

3.1 This technique is called SWOT analysis
In any new situation where it's not instantly self evident what should be done Think SWOT (see Fig 3.1).

S is for STRENGTHS
WHAT ARE OUR STRENGTHS IN THE SITUATION?
Yes, we do have a 2.1 degree in the relevant discipline,
Yes, we do have a healthy bank balance,
We are reasonably fluent in Spanish (thank you Uncle Miguel),
We do have a clean driving licence.

W is for WEAKNESSES
WHAT ARE OUR WEAKNESSES?
We do have small children to care for,
We are nearly 30 that being the required top age,
We have almost zero Word processing skill.

O is for OPPORTUNITIES
This is a chance to get into The Iberian Liaison Office,
If we score it could rate a Company Car.
One promotion can, if properly nurtured, lead on to others.

T is for THREATS
Can I really handle extra responsibility?
I might not get on so well with new people.

SWOT Analysis Fig 3.1

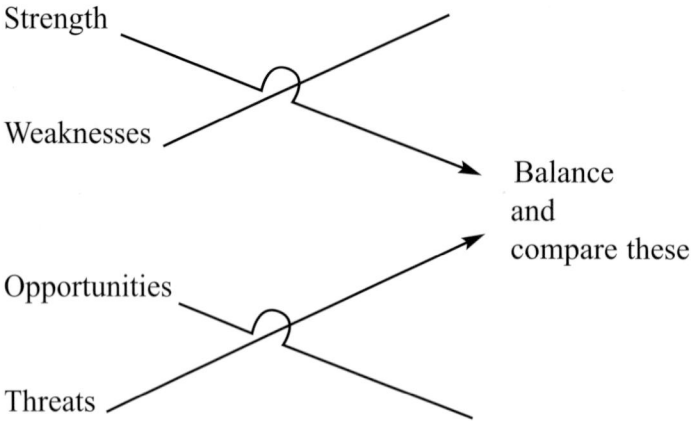

Strength

Weaknesses

Balance
and
compare these

Opportunities

Threats

The technique is widespread and old, and is available in set-out form which can be applied to many business situations, see SWOT on the internet if you wish to see the full range of offerings.

Whenever you do a SWOT analysis it serves to clarify the situation, especially if written down. The next step is an **action plan.**
This can range from 'do nothing' its best to leave things alone to seeing if a crash IT course is possible in time to be useful and then 'Go for it'.

3.2 ANIMALS IN THE OFFICE

Whatever your situation in modern times there always seems to be too much to do, too little time to do it. How do you prioritise? And sleep sound at night?

I'm going to introduce you to some animals who lurk in an engineer's office. Different engineers have different resident species but I'm sure you will be able to identify yours.
Look at Fig 3.2 there are four animals present.

3.2.1 The bluebottle about to drown in your coffee

not important, easy to spoon out and not urgent. Miss Jones can fix it in a trice!
Don't let bluebottles bother you.

Bluebottles = Not urgent and not important, so treat them as such.
Maybe old Fred is one – buzzes a lot but can be safely ignored.

3.2.2 Percy Parrot, forever, it seems rattling in your ear

Could be accounts deptartment wanting more and more monthly figures.
It's fairly important so best to set them out and shut the bird up.
Parrots have a certain urgency and are rather important so make some effort to fit them in.

Fig 3.2 Animals in the office

3.2.3 **The hooded cobra,** that's the direct phone line to the General manager's office. It could ring (strike) at any time at all, and commands instant attention.

The cobra is always urgent and very important – be very ready!

3.2.4 **Last we come to the sleeping dog, so peaceful and strokable** – surely he could not constitute a threat!

Not urgent but probably more important than the other three together.

What does the sleeping dog signify? What in your life is not pressing but if ignored for too long can rear up and bite – hard?

What do you know about progress towards Chartered Engineer status. Is a scheme in progress? Who is steering and guiding it? When a vacancy arises will CEng be in your CV or only hoped towards?
What is happening about continuing professional development? A couple of courses running or 'not as yet'?

You know there could be upgrade vacancies in the new Toulouse plant, do you keep up the French Language course?

Later we will be considering how to plan and execute by the day, week, month and year. As we work up to that start getting well used to recognising Bluebottles, Parrots, the Cobra and Sleeping dogs and getting into a routine for dealing with them.
Never forget – The Sleeping Dogs seem to be the most benign and ignorable, yet can in the long run hurt you most. Start early at bit-by-bit getting ready to greet them gladly when they wake up.

Questions

Are you now aware of what SWOT analysis is ?

Could you imagine applying it to purchase of a particular house or car?

Have you got a grip of the Animals in the Office?

Are you ready to treat them appropriately?

Chapter 4
Useful Development Techniques

● **Mind Maps**

● **Day Books, Daries**

● **Lists, Files.**

Work hard at laziness!

Let's look at some simple techniques useful to help progress through a busy existence.

4.1 A day book

It's well worth obtaining a hard-cover foolscap size book to act as a day-book (Fig 4.1) day events are recorded in brief scribble on a right hand page, appointments made, contacts made, promises made, names, phone numbers etc. Roughly how much time is spent doing what, how many miles got travelled (for car use claims) and the like. If you are working in a consultant mode coherent claims for time on various projects can be made. A new dated clean page for each day works well (Fig 4.2) This document can also act as a log book for purposes of effort towards Chartered Engineer status. Observations, later notes that relate etc. can go on the left hand page opposite to the day page.

I much prefer a hard-covered book in my brief case over electronic devices or a snowstorm of backs of envelopes. You choose, but do have some sort of non-volatile record.

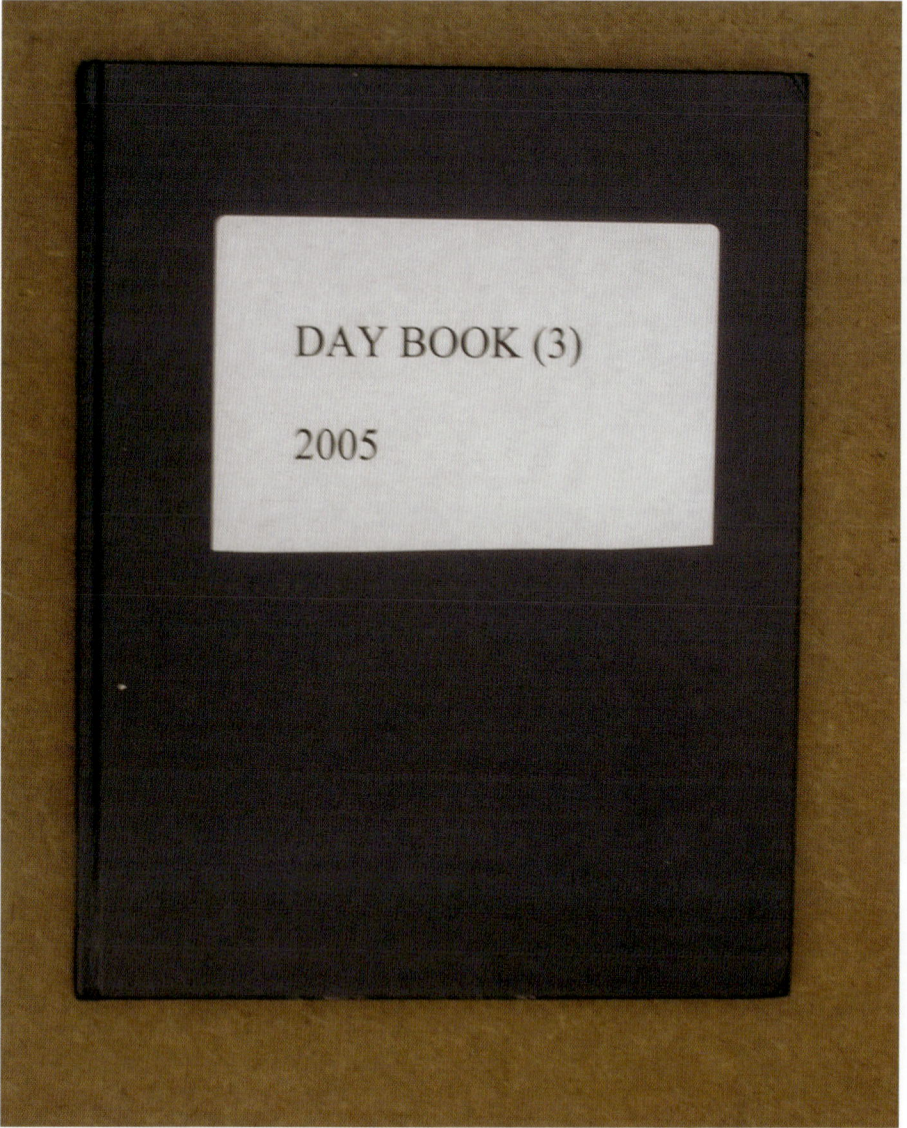

Fig 4.1 Day Book

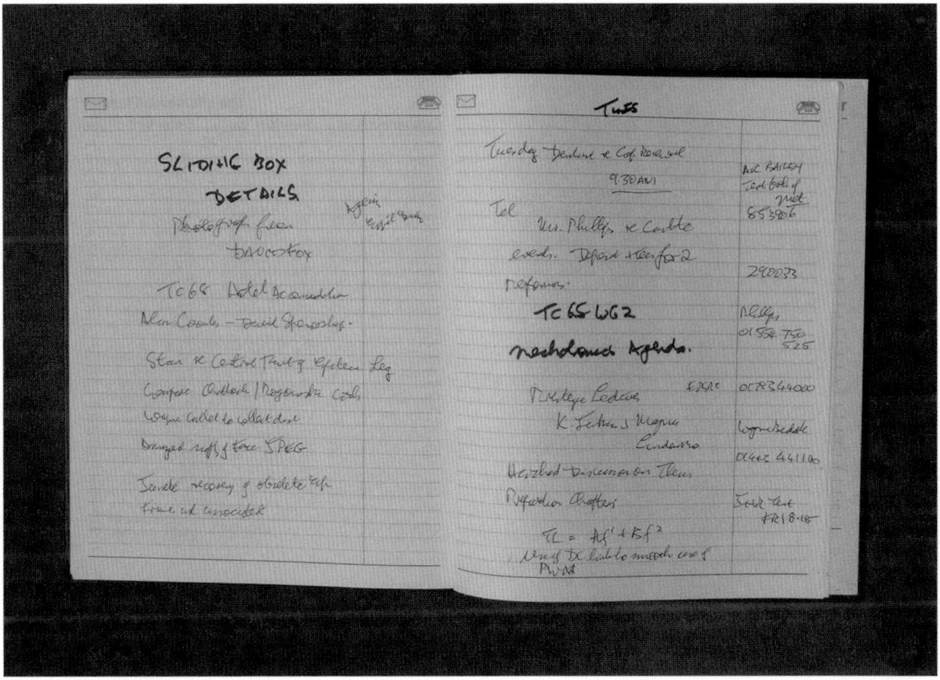

Fig 4.2 Day book pages

Even if you give way to the temptation to use it like a filing system – papers stuffed between relevant pages till it bulges (try not to) the scheme pays off when you need back information.

4.2 **The Diary (Fig 4.3)**

Most professional bodies (about whom we shall say more later) give out annual diaries.
On the early pages you can record personal details, your e-mail addresses etc. All the telephone numbers you need along with names and addresses laboriously copied out from last year are on hand. New ones may get scribbled in on the north sea, but are there somewhere. Could even be in the Morey Firth of the Map Page.

As appointments come up and are changed or added to pages and day spaces tend to solidify be it Dentist, Management meeting, submission of paper abstract last date for, cat-flu injection for Tiddles or whatever.

Many people try to convince me that an electronic device or space in a mobile phone memory will do as well or better. I however am sold on the ability to overscribble draw arrows etc on real paper.

Fig 4.3 Pocket Diary

On one woeful occasion I had scribbled a vital diagram and phone number on the back of a white corridor door. Soon this data was needed but alas my wife had beautifully repainted the door!

4.3 LISTS

There are list lovers and list haters.

Personally I keep running lists of :-

For the day:

AM, PM and Evening

For the week:

An outlook list

This latter covers pending matters and things that can forward ambitions or build achievement against the time when sleeping dogs will wake (they always do eventually).

Where you may ask is the opportunity for laziness?
Well if I see that all my evening 'jobs' are crossed off except three and two of those depend on mail that's not in yet and a third can wait till the boss is in a better mood, I can declare a mental holiday and watch inane TV, kiss my wife or simply put feet up and snooze (Fig 4.4).

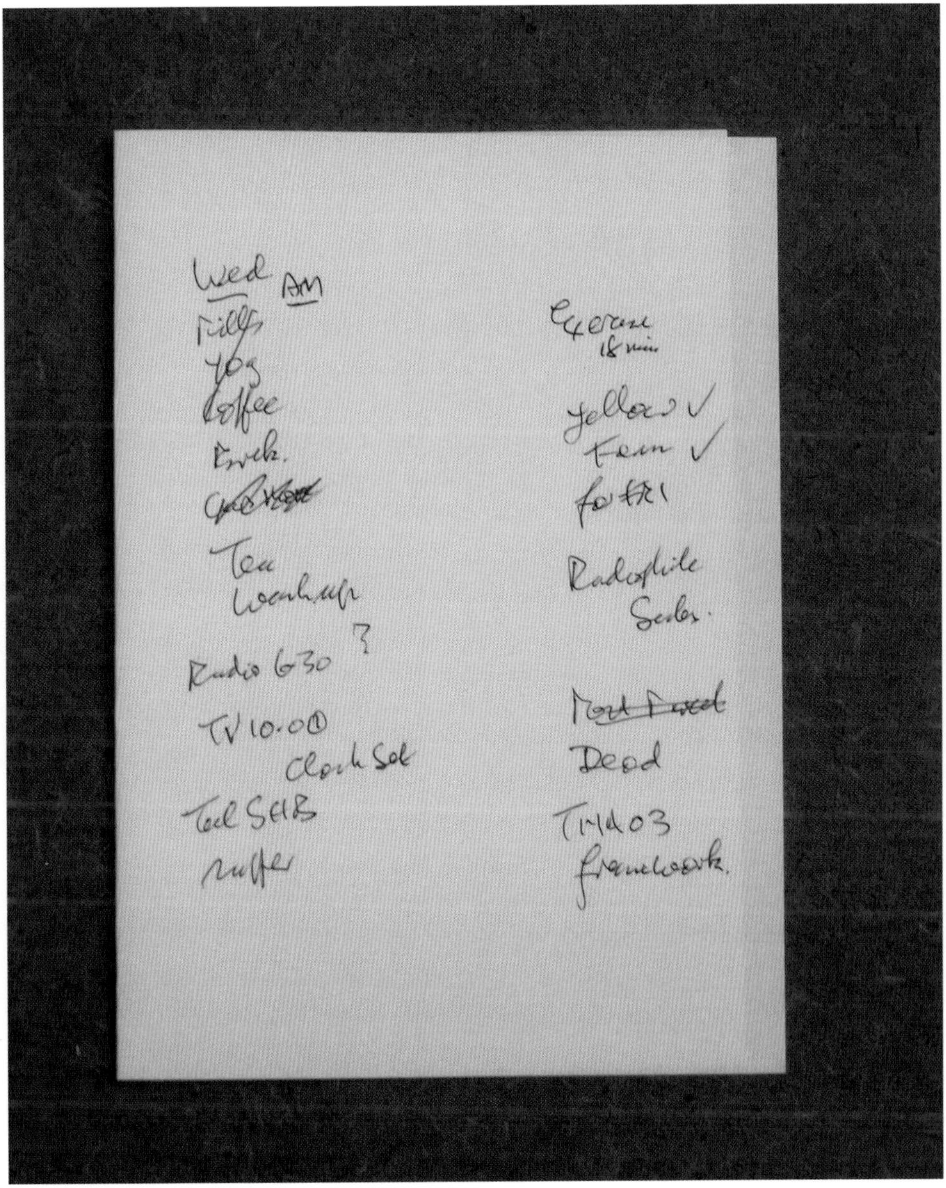

Fig 4.4 List Page

Instead of feeling pressured and rat-raced all dyke holes are well enough plugged to make glorious do-nothing legitimate – even mandatory.

The lesson is "use lists day-books and diaries to enable you to be effective as well as relaxed".

Going too far? Anything can be overdone if you are tempted to make lists of lists it's probably a sign of overkill.
If you have ever mislaid your car keys and looked 'everywhere' exploring the butter box in the fridge or the cats basket is unlikely to be productive – time to sit still and do a Sherlock Holmes act.

Figures 4.1, 4.2 , 4.3 and 4.4 give an idea of what your refined laziness documents could look like.

4.3 **FILES**

How much time do you spend hunting for pieces of paper? It's well worth while buying a quantity of poquet folders (different colours) and drifting things relating to an emerging project into a labelled one of these. If the projects splits into parts, eg. hotel booking, plane and travel arrangements, abstract of the paper you are to give etc. The poquet folders which relate can nest into a (labled!) box file. Eliminating much of the time usually spent hunting for and maybe losing papers contributes greatly to the fruits of benign laziness.
Engineering courses may be great at teaching integration by series but effective paper management has just as big a pay-off.

Most Departments have a member whose office is almost unenterable, unsteady piles of papers cover every chair,shelves cascade ancient journals onto an already layered floor. The top surface of his/her desk could accomodate hens in deep litter. Somewhere within the heap a phone rings plaintively.

We've all seen it, don't do it! Be ruthless with the little accretion, which is even now starting on the bedside table.

SO THERE YOU HAVE IT – WORK HARD AT LAZINESS!

Questions

Do you make appropriate use of a Daybook and Diary?

How do you prioritize your activities?

How do you create and defend 'for me' time?

Chapter 5
Publications

● **Papers**

● **Conferences**

● **Seminars**

● Animals in the Office

5.1 **Papers**

There is always plenty to do in a job without looking for more but   publication is one of the Sleeping–Dog issues which repays attention. You may not be working in a research Department but you will develop ideas and opinions about your field of activity.

By reading appropriate journals (some produced by and for the young engineer) it should soon become clear where there is a gap to be filled, or a letter to the Letter Page to be written.

The benefits are considerable, published papers look well in your CV and aid visibility.

Maybe a joint byline with your boss is appropriate, you can benefit from appearing in print in association with a senior figure. Of course relevant permission and acknowledgement have to be attended to. It's never too early to become even a little bit known in your field.

Sometimes employers' House Journals welcome interesting items especially the "what you always wanted to know but were afraid to ask" type.

A nodding acquaintance with journals such as New Scientist and Scientific American allows you to be interesting to talk to on nearly any topic.

At least it's one up on the agony letters in the ladies magazines found in the dentist's waiting room.

Publication in the most prestigeous journals is a lengthy and exacting process subject to strict peer review. Best to start lower down, maybe by submitting to a conference.

5.2 **Conferences**

We will consider in due course Learned Bodies. Many of these sponsor Conferences, workshops and local meetings.
Your organisation whether university or Industry can have a range of motivations for taking part, eg:-

a) Desire to be seen and have the department put in the public eye.
b) To keep a look out for potential employees.
c) Trawl for new ideas.

Submitting a paper or Poster
Well ahead of a conference there will be a call for Abstracts, probably a title and 100–200 words outlining the proposed paper or poster. The program Organiser will put together an outline program and if your Abstract is welcome may call for a full paper text. If you are well known in the field this latter request may be delayed till texts are submitted to peer review for publication

A good paper will be pitched at an appropriate technological level, neither patronising nor tending to smother with dense mathematics.
Clear and intelligible English is vital, combined with correct syntax and grammar. All too many authors are far from expert in writing. As may be appropriate, DO get a friend to proof read your material. Someone having English as Mother Tongue is especially needed to vet text produced by persons whose first language is not English.
Think of the poor paper reviewer who has maybe ten papers to vet in detail and pass for publication against a tight deadline.

Help him and secure a reputation for good communication.

Verbal presentations give you the opportunity to lay the foundations of a reputation for offerings that inform, even entertain.

Your techniques of effectively delivering a talk at a conference are best founded on experience gained in smaller Department seminars.

CONFERENCE POSTERS should be produced in accord with the requested size and format. If your Department's facilities can manage it, a single sheet of A0 paper on which a computer layout of your material has been assembled (with colour) looks very well.

Problem: A rolled up A0 sheet is not easy to protect through air travel to a foreign conference.

Remember, **Business Cards** ready to hand at the poster site do good advertising for you and can be deposited at other poster sites to request pre-prints of papers.

In any case avoid putting up a poster which consists of a couple of dog-eared pages drawing pinned in place.

5.3 **Seminars**

Seminars consisting of three or four 10 or 20 minute talks followed by discussion are often put on by Company Departments or University Engineering Schools.

Often speakers have to be press ganged into willingness to appear. Don't make this mistake, volunteer. Seminars are a very useful medium for letting you become visible and known.

Many publications exist offering advice and training in "Public Speaking." We will say more about this later.

Questions

Do you look for opportunities to deliver items at Seminars?

Do you work up papers relevant to your specialism for publication?

Do you attend conferences?

Do you help in the organisation or mechanics of getting a conference to go in your area?

Chapter 6

- **The Engineering Community**

- **Membership of Professional Bodies**

- **Networking**

- **Officer, Chairpersons**

Membership of Engineering Bodies

The Engineering Council is made up of some 30+ members such as The Institution of Engineering and Technology (formerly Electrical Engineers), The institution of Mechanical Engineers and so on (see Appendix).

These bodies have regional associated structures within which regional activities take place. Typically lecture evenings, industrial visits and social events make up a programme for each year. Some events are set up particularly for younger and student members.

From time to time Engineering Institutions will visit universities and hold recruitment fairs aimed at recruiting new and student members.

There is much to be gained from membership of one or more such bodies. The Engineering journals which members receive keep them informed about current issues both technical and political. Attendance at regional meetings provides invaluable opportunities for meeting people over the inevitable tea and biscuits.

This process known as **NETWORKING** is very important, giving young engineers present the chance to mingle socially with senior figures. Even if a lecture was on a topic nowhere near one's own specialism it is rare to come away without having made a contact of real use or interest.

NETWORKING happens at local engineering meetings, learned body events, conferences, Trade Fairs and many more.

Effort expended in learning how to meet new people is never wasted, it promotes ease of manner and raises visibility.

Engineering bodies are always in need of help. From time to time a new Secretary, Treasurer or the like is being sought. It's well worth volunteering for such a post, often constitutions limit a post to two years for any one person, likewise Committee members.

Such engagements open up more networking as ex officio you then <u>have to</u> meet people.

CHAIRPERSONS at some time or another you are likely to be deputed to chair a seminar. DO IT WELL.

We have all seen the shambles that occurs from time to time

Key points are:-

a) Visit the room and get it unlocked ahead of time, avoid embarrassing delays while the key is sought or pirate occupants are evicted.

b) Ensure that overhead projector, video projector and white board (black in times past) are in place and working, as may be needed. Have a laser or other pointer present

c) Find the Speakers, be certain of their names (spell), affiliation background and topic.

d) Agree with speakers about time limits and how you will stop the inveterate overrunner (custard pie?).

e) Listen to the presentations and be ready to frame a question of your own if no-one else has any.

f) Have someone primed to give a vote of thanks etc.

g) Ensure that if tea/coffee is on offer it is delivered ahead of or after presentations. No speaker likes to battle with the sound and bustle of crockery being set out.

Questions

Do you seek openings for Networking?

Do you actively enjoy meeting people?

Do you realise that 99 % of an audience wishes the speaker well and mentally urge him/her on to give a good successful talk?

Chapter 7
Image Projection and Referees

7.1 Interviews

We have seen that the Car Park and related tests take note of Image and Impression.

The Formal Interview is a much used tool for selection and hiring. Many books deal with interviews and interview technique. Examination of some of these is useful, but by far the most useful training comes from "meeting people". This can come about by participation in Engineering Society and the consequential networking. Trying to commit to memory things to do and say is seldom completely successful. Far better to be ones you have developed over dozens of informal networking situations. However Image Projection is important. I was once disappointed to find that a candidate, whom we did hire, did well at interview and looked very good, but when reporting for work on the first day of employment had 'lost' the crisp interview attire and smart appearance and presented in jeans, T shirt and designer stubble. Subsequent work performance was equally variable!
In modern times it is strictly forbidden to comment on details of personal appearance, still less to impose rules about it.

None the less men and women do make unspoken statements about themselves, which, like it or not, do lodge in the memories of their employers and can influence the view through the Car Park window.
It's all very fine to claim rights of appearance, but Engineers are responsible for the effects of the impressions which they give A baseball hat worn indoors (back to front?) or a bone through the nose may be your inalienable right, but if the manager is going to call someone out of the depths of the development laboratory to help a customer – whom will he call?

Can you cope with bizarre situations and not panic?

I well remember travelling to Sweden to interview a candidate for a senior position. Unfortunately there was muddle over the booking of a room at Stockholm's Arlanda airport and no time in hand as I was due back in London via the next flight.
We were routed through endless underground tunnels under the guidance (?) of an embarrassed booking agent, then wrong room – no key!
Back again via more tunnels – mind the builders bricks and cables!
In fact I interviewed the man on the move in semi-darkness and could take no notes. As it happened he stood up to this situation very well, we hired him and he is now, years later, a very senior indispensable manager.

One of the mental checks which I apply is to ask myself the question, "If marooned on a desert island would this person be a great support or an additional drag on escape attempts?"
Adaptability and lateral thinking are not readily taught. They spring from an ongoing need for them in your life.

I am usually impressed when people reveal an interest in demanding hobbies.
An enthusiasm for Aeromodelling, Amateur Radio, Old Clock restoration or the like usually conceals the fact that that there has been need for careful budgeting, lateral thinking, improvisation, persistence, attention to detail and so on.
All excellent self-training for becoming an effective engineer.

Sport deserves a mention.
Participation in sport develops many good qualities. I well remember turning up for a Marathon run having just found that my car had been broken in to and items stolen.

At about 20 miles I was getting more and more weary, but the thought of and anger over the car break-in fed adrenaline into my system when most needed and I made a fair finish.
Later on my way home, when reporting theft to the police they looked me over and said "why did I come in such funny clothes?"

Business Cards, as mentioned in Chapter 2, are an item to ponder.

It's very useful to be able to leave your persona in the hand of the Networker after the cup of tea. Get remembered and called on when an effective engineer is wanted!

Referees

From time to time in your career you will be asked to provide the names of referees.
It may be for a change of grade in an Engineering Institution, or it may be for a job application, even a Passport application.
Memo make sure you have a valid passport and could go anyplace in a hurry on behalf of your Company.

Networking and rubbing shoulders with the core people in your field gradually builds up a 'bank balance' of persons able and more importantly willing to act as a referee for you.
When you become old and grey, (bald?) remember those who helped you and lend a hand to the promising lad or lass whom you may know.

Chapter 8
Aspects of the mind

8.1 Sociability and the Mind

Besides being a business unit an Engineering Company or University Department is an assembly of people.
They have hopes, fears, joys and sorrows, as do we all.
Each face we meet and speak to has a person behind it.
We may only be aware of the workplace front.
My time in hospital taught me that for sure 'no one person is an island' we are all joined into human society. I further learned that many of the Doctors and Nurses carried heavy unspoken loads on their backs, yet made time to care for me and be supportive or even issue reprimands when it helped. As a result I now see Engineers a little differently, including those with disabled children, dying wives and so on. I am more able to look at the face behind the public face and try to support the whole person where I can.

When one keen young engineer was eager to deliver data to me on time I suggested that I pick it up from his home. I did so. I was amazed to see the poverty in which he lived. 10p tins of beans with last weeks stale bread and so on. I had not realised that he was getting by on almost zero finances, but keen to do well, the work never suffered and his turn-out was always clean and tidy.

This spurred me into researching ways to find a small extra bursary for him by squeezing some already near-dry budgets. I'm very glad I did, he completed his research very creditably and is now years on occupying a respected managerial post in Industry.
There are more people struggling bravely and keeping it hidden than we know about.

This forces me to remember that when I was a secondary school pupil a local timber merchant would buy me one or two Physics books each year that I could not then aspire to afford.

Dead now alas, but the signed copies remain with me.

As an Engineer, avoid being exploited, but where it counts try to help lame dogs over stiles. They may be Corporation chairpersons one day and offer you a job!

8.2 **The Theatre of the Mind**

To do well at a task it is invaluable to be able to practice for it.

There is a long-standing notion particularly in the USA that we can play out rôles in the Theatre of the Mind (see many management books). Sitting quietly we create a vision of the situation in which we have to put forward the case for a new type of plant to the Company Board. Have we prepared the arguments? Do we know the relevant facts? Can we envisage getting up and explaining it all? Practicing it in the Theatre of our Mind can prepare us for the next stages of our job or the inevitable encounters coming up in our current one.

Can we see ourselves and (hear ourselves) getting by in Japanese or French- will we try harder with the language lessons now? (Sleeping Dogs).

8.3 **Body Language**

I have heard at least one person's walking described as 'he falls along'. It's much better to cultivate a steady deportment and appear to be well joined up at the corners. Since undergoing physiotherapy I have been coaxed to discipline lax muscles and write new cerebral software to operate those left unprogrammed.

8.4 **Handshakes**

Handshakes are such an everyday exchange of contacts yet they do much to make statements about us.

We are not inspired by the LIMP FISH.

The HOT AND SWEATY feels evasive.

The BONECRUSHER we suspect of an overdeveloped ego.

A good Engineer we hope to be COOL AND FIRM and not overprolonged.

We appreciate the toil that has procured the coarsened palm of the Landscape Gardener.
Once again its like the Car Park, nowt said but all noticed.

8.5 Speech

Speech is a tricky one.
I was lucky enough to have been exposed to a Classical education and find the discipline of dead languages' an automatic help towards accurate speech and syntax.
Likewise I have been shuffled around Europe enough to find familiarity with their modern descendants.
For business purposes I have had to learn enough Japanese to hold my end up in performing magic tricks and singing Japanese songs in Tokyo night clubs.
Hospitals have many immigrant nurses these days and it is surprising how a vocabulary of some 5 words and phrases in, say, Tagulog bring out smiles and empathy.
Step for a moment into another's shoes and help him/her to smile!

So much for language, but I failed on regional accent. Being a native of darkest Gloucestershire my voice seems to me to be typical of a yokel leaning over a gate chewing a straw.

I have never fought to change it much because as soon as Engineering becomes the topic it seems irrelevant.

In Wales I am an immigrant from England. After 40 years there they almost seem to think I'm human now!

8.6 **Office Social Occasions**

The aim of Office Parties and Outings is everyone's enjoyment.
The Young Engineer, Survivor of the Car Park Test needs to realise that the office event can become the Party Test as you are likely to make statements about yourself.
The clever trick is to have a jolly good time and retain a top line image.
"He's fun – we must ask him to ....event".

8.7 **Mind Maps**

No one's memory is perfect and mine resembles a sieve.
If presenting a Seminar or tabling a scheme at a meeting or writing a complex report a Mind Map can be invaluable.
Figs 8.1/2/3 show a map being developed for a seminar item.
The Subject goes first at the centre then consequential ideas radiate from it, re-branch and then, maybe, interlock.
Coloured versions are considered to be even more powerful.

The creation of the map.
This engraves the logic of its creation in the brain and greatly eases its reproduction as may be required.
This can be a helpful route to developing a new product or device or problem solving.
I suggest you may like to consult 'The Mindmap Book' by Tony Buzan (see bibliography).

Fig 8.1 Development of a Mind Map
Topic – Waveform Distortion

Step 1 put topic centre-page :-

Waveform Distortion

Fig 8.1 Step 2 - add considerations arising from the topic

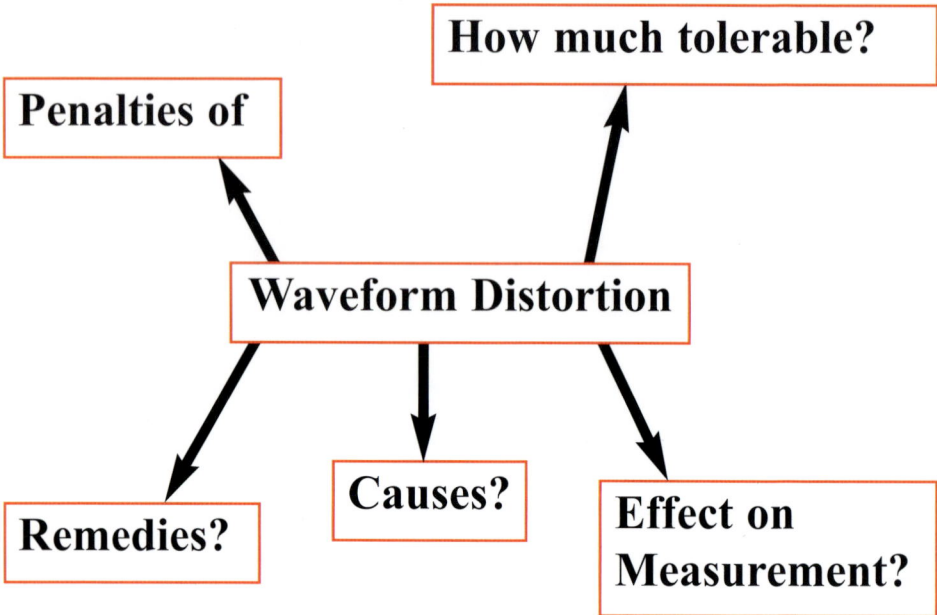

Fig 8.1 Step 3 Ramifications.

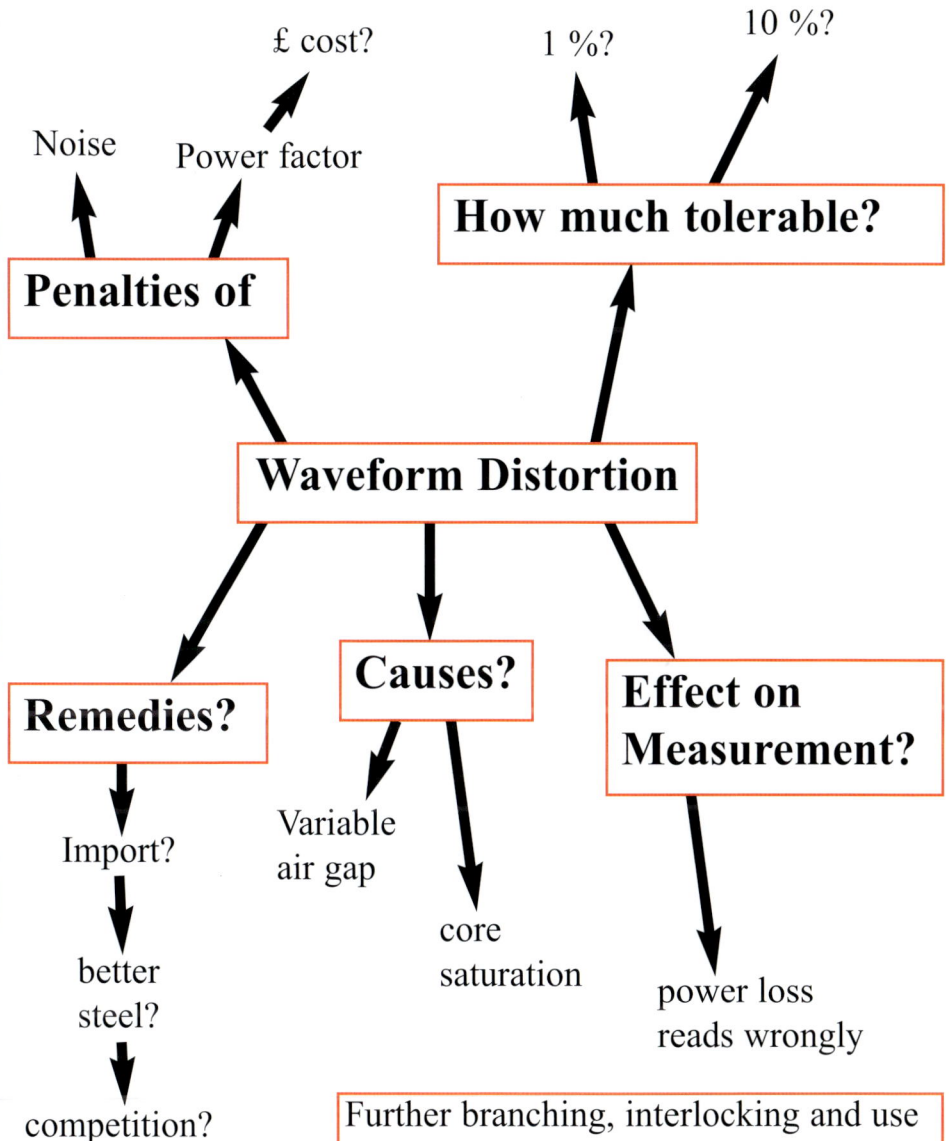

£ cost?

1 %?

10 %?

Noise   Power factor

Penalties of

How much tolerable?

Waveform Distortion

Remedies?   **Causes?**   **Effect on Measurement?**

Import?   Variable air gap

better steel?   core saturation   power loss reads wrongly

competition?

Further branching, interlocking and use of colour all aid usefulness.
See Buzan's book (Bibliography).

8.8 **Self Help**

The reader will recall that we have been examining and considering the place of Cement-Type Training and Education.

The whole area of self-help is very much with us these days.

It is not necessarily relevant to embark on a whole MBA course, but a few things yield good fruit.

Although now old, the book 'How to win friends and influence people' by Dale Carnegie still offers a useful read (see bibliography).

Also 'The One Minute Manager' (see bibliography) may be just as useful for your job as for keeping your children busy.

8.9 **Odd points**

You have seen me in this text carefully skirting round the edges of political correctness.

The latest to confront me has been the official requirement that Brainstorming sessions are not held any more. They have become 'Idea Showers', lest we offend victims of cerebral disturbance and the like.

8.10 **Memorableness**

It is sometimes entertaining to be reminded about things from the past.

Long ago I lectured to metallurgists on Electrical Engineering which their course required.

Some had travelled 150 miles after a days work to get to an evening class, after which they were due back on a night shift. Thus my lectures had to be a little dramatic to keep them **awake**. Recently, I came round the end of an aisle in the local Supermarket face-to-face with student XXX. Apparently forgetting a two decade gap, he said "Sorry I couldn't hand in your last homework, I got stuck on the calculations!". You know dedication when you meet it.

Some years ago I performed a demonstration of a matter of electromagnetism, which involved my suspension upside down by magnetic attraction on steel-shod boots.
I fear that the hang up will be remembered long after the equations have faded from the audience's mind while I was undergoing treatment a Physiotherapist enquired.
"You'll be strengthening muscles for the Boots then?"
How did she know?

Chapter 9
Rings of Capability

Getting an overall view of ones capabilities is a useful step towards creating an Action Plan and Road Map for your career.

What do I mean by a Ring of Capability?

Most Engineers are familiar with Polar Plots of quantities such as material modulus or strength.

In those shown here the distance from the diagram centre to the defining line or ultimately to the circumference is a measure of capability. Line length is scalar, the angle which makes it a vector relates to the property expressed.

The position of this vector and the clutch of directions near it define the subject group of capabilities (See Fig 9.1.).

Capability can range from zero to, say, 10 (all the way to the periphery) and to represent mathematical capability it could look like Fig 9.2 for a trained engineer. For an 'A' Level contender it is more like Fig 9.3. The maths is narrower in scope and lower in extent.

Extending this sort of plot to the full circle of capabilities the young student may look like Fig 9.4 – some of everything but not huge amounts.

The capability footprint for an older experienced engineer expands as in Fig 9.5, however the Life Skills lobe is not very developed.

The experienced, successful engineering manager has a larger and more impressive footprint.

I would draw your attention in this book to the importance of enhancing the Life Skill aspect of your footprint.

Figure 9.1 **Capability Ring**

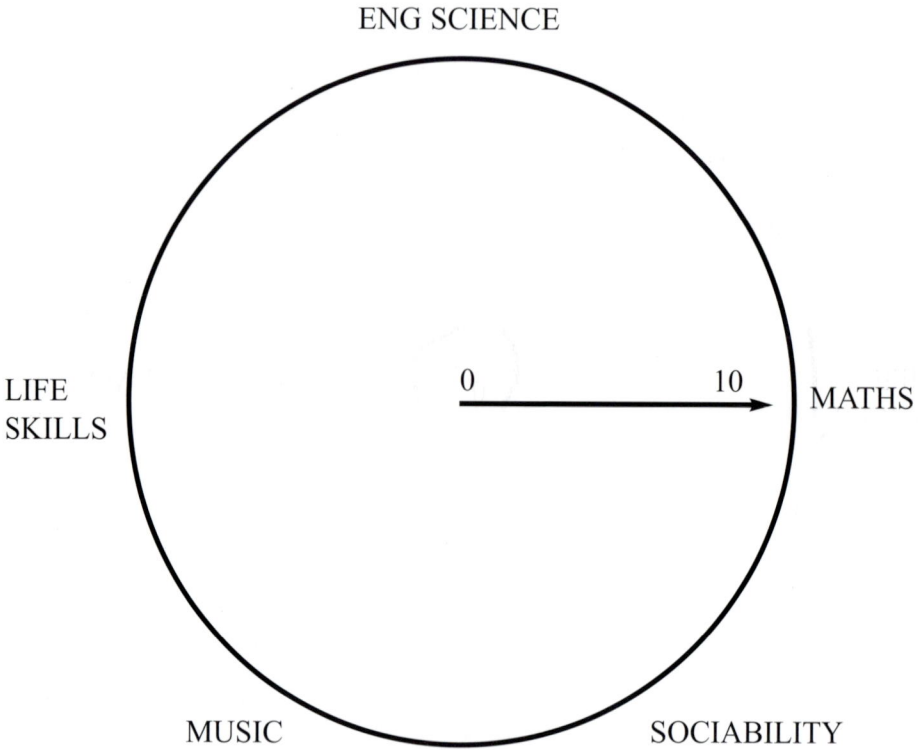

ENG SCIENCE

LIFE
SKILLS

0   10   MATHS

MUSIC   SOCIABILITY

Depth of capability on a scale 0-10 (centre-to-rim)
A person is described by a spread of vectors round the circle.
Notional capability regions shown here.

Figure 9.2

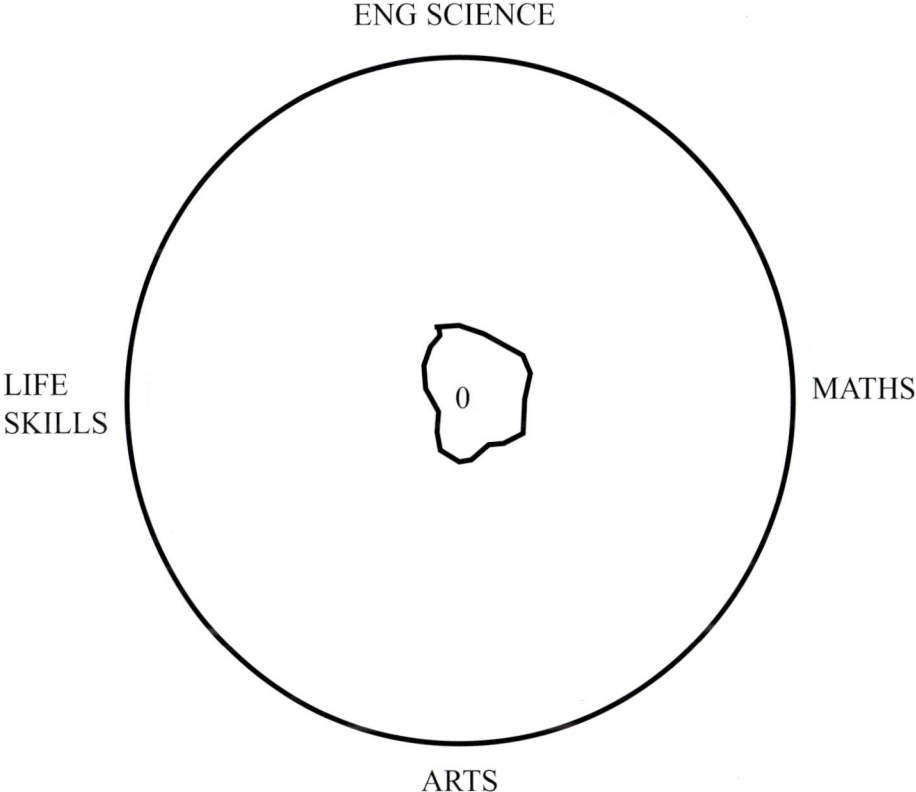

ENG SCIENCE

LIFE
SKILLS

MATHS

0

ARTS

Capability footprint for an "A" level candidate

Figure 9.3

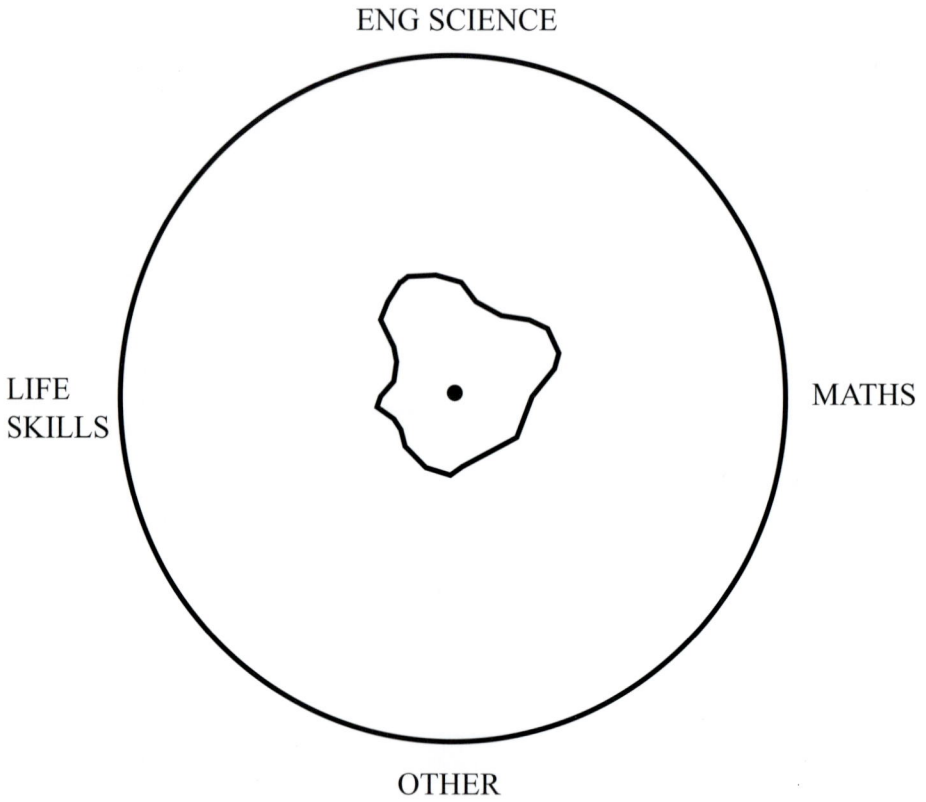

ENG SCIENCE

LIFE
SKILLS

MATHS

OTHER

Footprint for a graduate engineer

Figure 9.4

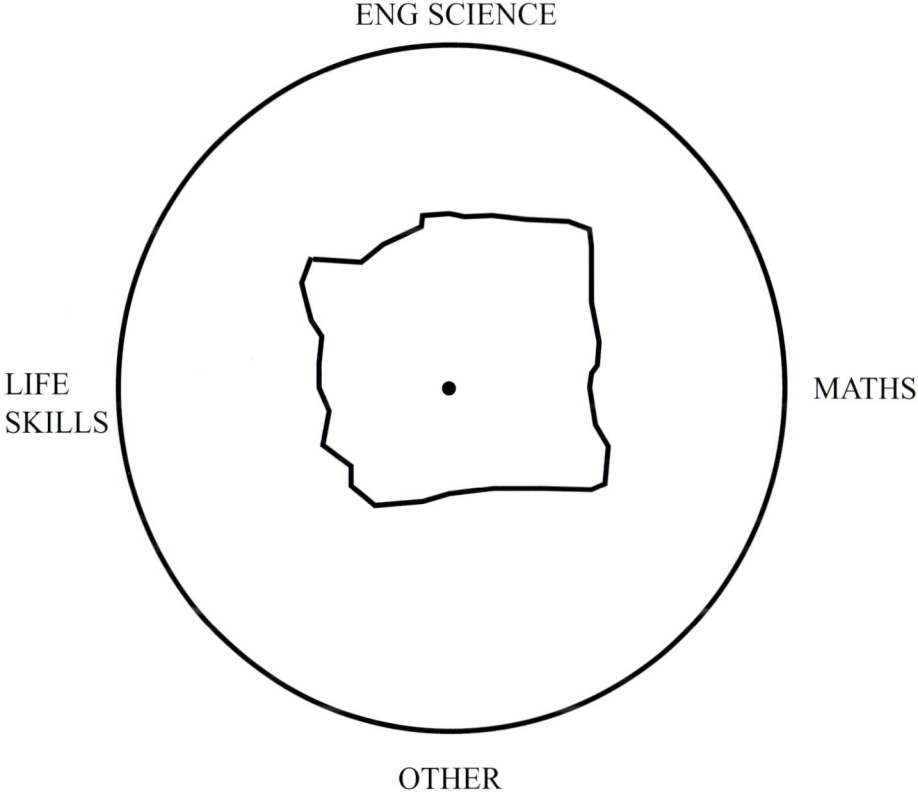

ENG SCIENCE

LIFE
SKILLS

MATHS

OTHER

Footprint for a fully qualified engineer with experience

Figure 9.5

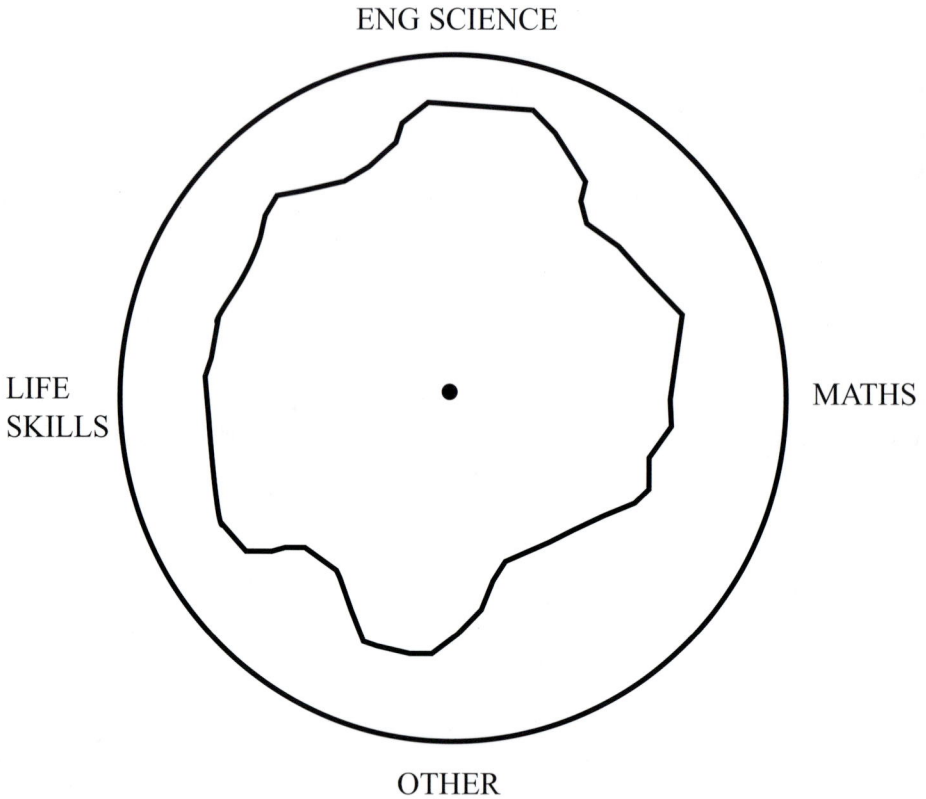

ENG SCIENCE

LIFE
SKILLS

MATHS

OTHER

Footprint for an experienced successful engineer manager - an
impressive footprint. An Engineer should aim to push the
envelope wider as training and experience grow.

In recognising the preponderance of hard-core training and the shortage of 'cement' training the Engineer can imagine attitudes and activities aimed at expanding his/her capability footprint towards the shape and size which scores well.

Of course no-one ever fills the circle but awareness of the need to try develops over time a very useful print enhancement.

A limited competence in a few less usual areas can bring good returns. For instance an overall grasp of Intellectual Property matters (IP), Patents and patent law contributes to the writing and discussion of papers in gestation.

On quite a different tack a colleague may have an excellent singing voice or a way with Monologues, all nice to have around in the later stages of e.g. Official Dinners.

When visiting a factory Engineers ask "how does it work? what material is used here etc. etc.?"

The visiting CEO asks "how many men do you employ? what is the shape of your supply chain?" or, "where are Company shares in relation to others, are they publicly traded etc? do you outperform the Footsie? do workers have access to share bonuses etc?"

Engineers may not need to be financial wizards but a an awareness of how it all works facilitates your visibility and aids conversation. One can even imagine a "visiting a factory with the CEO" test!

Questions

Have you got a vision in mind of the Footprint you have in your ring of capability?

Are you enlarging it? Do you know how to?

Have you got an outline of Intellectual Property matters in your head?

Can you ask intelligent broad financial questions about a rival's plant if visiting?

Chapter 10
Questionnarie and Interviews

10.1 **Preamble**

We all have views about Engineers, some our own, some derived from various arms of the Public Press.
Do we know what others really think?
To address this question I have devised a questionnaire to be put to some 20 Senior Engineers aimed at assessing their views.
No names are disclosed so there is no chance of recriminations.
Additionally I am seeking to interview the same persons and obtain some reaction from comparison of their outlook with mine and with the prevailing trend of opinion in Engineering.

10.2 **THE QUESTIONNAIRE**

The questionnaire used was.

1). On a scale 1–10, where 10 is the most valuable, how do you rate consistent punctuality in an Engineer?

2). How much importance do you attach to ability with Foreign Languages and an interest in these? (1-10).
a) European,
b) Japanese / Chinese / Russian?

3). How useful is an attitude of helpfulness, but not a doormat? (1-10).

4). How much importance do you attach to publication of Papers / Articles? (1-10).

5). Do you think attendance at Conferences / Seminars useful? (1-10).

6). How important is an Engineer's ability to play the rôle of Front-man? (1-10).

7). How important is it for an Engineer to be able to talk well? (1-10).

8). How important is it for an Engineer to be sociable but not fawning? (1-10).

9). How important is it for an Engineers to 'make' their own 'luck'? (1-10).

10). How important is it for an Engineer to fulfil a job in the face of hang-ups in other areas – e.g. deliveries? (1-10).

11). How important is it for an Engineer to be able to be a Showman for your Organisation, and be sent to cover a Customer problem? (1-10).

12). How important is it for an Engineer to return 'phone calls as promised?

13). LUCKILY OMITTED).

14). How important is it for Engineers to be almost never 'sick'? (1-10).

15). How important is it for an Engineer, if need be, to play a 'waiting game'?
(1-10).

16). How important is it for an Engineer to have or have had Hobbies which do their own teaching on how to budget time/cash etc.?
(1-10).

17). How useful may a 'good' Engineer be on a Desert Island?
(1-10).

18). What attributes in an Engineer impress you and cause you to wish he/she was part of your outfit? Apart that is from their taught Degree?

Name 3: [a,   b,   c.   ]

19). Do you perceive universities teaching Life Skills?
Can they be taught , or are they innate?

Comment: [_____

_____

_____]

20). do you think written reports need more attention?
Tick: Yes   No.   In general adequate.

I would be willing to take part in a ten minute interview

Yes   No

Name   _____

Chapter 11

11.1 The results of the Questionnaire are as set out below

The Questionnaire asked for a rating in the range 1-10 for each question. After some consideration it was decided to record the Modal number for each rather than the mean or median. This was done and fortunately there were no signs of significant bi-modal responses.
These results are shown in the form of a completed questionnaire with filled in boxes.
The attributes sought, as listed in replies to Q18 may be summarised as :-
PRESENTATION SKILLS, ANALYTICAL CAPABILITY, VERSATILE, PERSISTENT, RELIABLE, WILLING TO LEAD, ENG-RELATED HOBBIES, GOOD COMMUNICATOR.

This is very much what may be looked for in anyone within an organisation.

Question 19 produced responses, which gave the impression that Innate capability, hard experience and organised teaching all contribute to overall capacity, but that much more could be gained from teaching if this were to be applied more widely and curricular space made for it.

Question 20. Almost without exception questionnaire completers concluded that written reports needed more structured teaching and guidance.
In general the scope and direction of this book appears to fit well with the aspirations of stakeholders.
Without delving into the complexities of psychometric measurements this questionnaire has proved to be useful and informative. Readers intent on shaping their skills could well compare themselves to its modal results.

QUESTIONAIRE AND INTERVIEWS (showing responses with modal figures)

11.2 Preamble

We all have views about Engineers, some our own, some derived from various arms of the Public Press.
Do we know what others really think?
To address this question I have devised a questionnaire to be put to some 20 older Engineers aimed at assessing their views.
 No names are disclosed so there is no chance of recriminations.
Additionally I am seeking to interview the same persons and obtain some reaction from comparison of their outlook with mine and with the prevailing trend of opinion in Engineering.

11.3 THE QUESTIONNAIRE

1). On a scale 1–10, where 10 is the most valuable, how do you rate consistent punctuality in an Engineer?   `8.5`

2). How much importance do you attach to ability with Foreign Languages and an interest in these? (1-10).
a) European,   `7`
b) Japanese / Chinese / Russian.?   `3.5`

3). How useful is an attitude of helpfulness, but not a doormat? (1-10).   `8`

4). How much importance do you attach to publication of Papers/Articles? (1-10).

| 8 |

5). Do you think attendance at Conferences / Seminars useful? (1-10).

| 8 |

6). How important is an Engineer's ability to play the rôle of Front-man? (1-10).

| 9 |

7). How important is it for an Engineer to be able to talk well? (1-10).

| 9 |

8). How important is it for an Engineer to be sociable but not fawning? (1-10).

| 7 |

9). How important is it for an Engineers to 'make' their own 'luck'? (1-10).

| 9.5 |

10). How important is it for an Engineer to fulfil a job in the face of hang-ups in other areas – e.g. deliveries? (1-10)

| 9 |

11). How important is it for an Engineer to be able to be a Showman for your Organisation, and be sent to cover a Customer problem? (1-10).

| 8.5 |

12). How important is it for an Engineer to return 'phone calls as promised?'

| 10 |

14). How important is it for Engineers to be almost never 'sick'? (1-10).

| 7 |

15). How important is it for an Engineer, if need be, to play a 'waiting game'? (1-10).

| 6.5 |

16). How important is it for an Engineer to have or have had Hobbies which do their own teaching on how to budget time/cash etc?(1-10).

| 7 |

17). How useful may a 'good' Engineer be on a Desert Island? (1-10).

| 10 |

18). What attributes in an Engineer impress you and cause you to wish he/she was part of your outfit? Apart that is from their taught Degree?
See chapter 11

Name 3: [a,   b,   c.   ]

19). Do you perceive universities teaching Life Skills?
Can they be taught, or are they innate?
See chapter 11

Comment: [

                                                                    ]

20). Do you think written reports need more attention? Tick: Yes,   No.
In general adequate.
See chapter 11

I would be willing to take part in a ten minute interview

Yes   No

11.4 **RINGS OF CAPABILITY**

Where questionnaire completers chose to pass comment on the (purely notional) capability footprints of Chapter 9 the feeling was that mathematics capability peaks during studentship and decays thereafter. Eminent Persons please note!!.

11.5 **INTERVIEWS**

Scope of Interviews

●Seven persons were interviewed of Status:-
●Technical Director (eng materials commodity industry)
●Manager Technical Laboratory
●Senior Manager Tech/R+D
●Professor (magnetic Materials)
●Director (University engineering School)
●Technical Director (manufacturing industry)
●Post Doctoral research Fellow

Discussion centred round questions :-
1. Sources of your life skills insight – Innate, Experience, Taught?
2. Reaching you via written media? or people?
3. Cross-Cultural insights (foreign and Industry/Academia), Training or pushed in at deep end only?
4. Involvements like Chairperson Secretary etc of bodies outside of employment.
5. Should life-skill training be incorporated in what are claimed to be already heavily overloaded syllabi or reliance placed on the 'in at the deep end' and learn by experience method?

Digest of Outcomes of Interviews

a) In all cases life skills were developed as a mix of Innate ability, Experience and some taught courses.

b) Contact with people has been more influential than material conveyed on paper.

c) Little training is given on cross-boundary operating, either foreign or Industry/Academia.

   Trial and error can be painful, but once learned lasts a lifetime.

d) Publications and Conference attendance pay off in terms of networking.

e) There is a definite belief that The UK needs life-style training of younger engineers to support and maintain cutting edges in Industry and Academia. Lack of time and budget are mentioned but payback is accepted as real and valuable.

f) Engineers are prompted to hold the initiative and LEAD.

Chapter 12

Analysis of the Questionnaire and Interview responses showed that
Employers want a whole range of useful traits.
Have you got these qualities? Is it reasonable to expect them?
What are you doing to add 'Cement' to Hardcore to firm up your
prospects?
The predominant opinion quoted was that some life skills are innate, some
arise from experience and if taught can tip the balance favourably.
In some quarters "Self Improvement" is a despised concept, but evidence
so far suggests that beyond grasping whatever training may be offered a
positive search for routes to excellence should be made.

Chapter 13

Action Plan / Road Map.

If you want to mix useful cement, consider the steps below:

1) Can you do a SWOT analysis on yourself?

2) What Strengths can I easily exploit or add to?

3) What Weaknesses can I begin to compensate for?

4) What might hold me back?

Questions

How many Sleeping Dogs are under my desk Can I deal with some of them?

Can I improve my Language skills?

Can I address and manage Seminars?

Can I publish some of my work?

Am I interesting to talk with?

Do I suffer from Halitosis? – please double check and fix it.

What useful books can I read?

Can I enjoy all this rather than seeing it as a burden?

After all you managed an Engineering Degree so it should be straightforward.

Am I doing the appropriate thing to move towards Cement mixing?

Finally, I'm glad I opted for Engineering – I really do hope you will be also.

Envoi
Innovation

While keeping the main thrust of Life Skills development in mind it is useful to remember that in a knowledge based industry, as most are becoming, the ability to innovate is a valuable capability for selling ones usefulness.

More and more engineers are being expected to think 'out of the box'. Setbacks and failures can be seen as opportunities to show innovative and recovery skills.

A failure can stir up ones thoughts towards productive action – 'Stimulate to Innovate' Schrage has said.

Often opportunities for innovation pass under ones nose unnoticed.

A few examples:-

a) While lecturing in Radar Techniques (50 years ago) in the Air Force we found that steak pies placed in the laboratory waveguide would warm nicely while doubling as a dummy load for the high power magnetron and deliver a delightful odour.

Regrettably we worried about being detected in illicit snacks and failed to imagine developing a microwave oven industry.

b) While developing a flaw detector for tin-plate it became problematical to keep return-current magnetising conductors out of the main observation area.

Jokingly I said 'locate them at infinity'. Quite soon it was found that for the scale of our work one metre gave a good approximation to infinity. The practice became the basis of a patent and a profitable commercial device.

c) Field sensing conductors encircling steel strip were easily damaged and this led to a lot of equipment down time. In fact a fine jet of dilute ammonia solution was found to serve as well as copper with the advantage of being non-hazardous and instantly self repairing.

The PhD Thesis

Project supervisors are often so closely enmeshed in a student's PhD project that little thought is given to the External Examiner.
When offering mentoring support to PhD students it is vital to have hammered out a Strategy for the job.
What is it about?
What are its aims?
Probable intention areas may be :-
To uncover new knowledge,
To develop a workable theory about XX,
To bring phenomenon YY from obscurity up to a useful prototype application stage,
To extend the range of material compositions, temperatures or whatever within which phenomenon ZZ has coherence.

So its vital to be clear what animal is being hunted and what for.
After a Strategy session has pinned down the essential aim consideration can be given to ways and means.
This can take the form of :-
a) A literature survey and a review of what has gone before, other workers, key methods and their findings.
b) What experiments are to be mounted? How may data be accumulated?
c) Does expectation require development of New Techniques? – are old ones to hand and useable? What materials/services will be required? At what cost £ $ € ?
d) Probable calendar period for experimentation and data acquiry?
e) 'How is it going' sessions each two months or so, asking 'is this project still viable, have we to change track radically?
f) what should be the last data taking day?
   (I vividly recall trekking to the Lab in the middle of the night to capture just one more instrument reading).
h) Thesis writing time plan, last allowed date?

Overall Map

You should end up your forward planning with:-
Former work list and papers review,
Experimental programme,
Validity of data,
Methods used,
Display of data,
Discussion of results,
One sentence statement of 'were aims achieved' ?
Formal conclusions,
Outlook for further work.

THE MESSAGE IS:-
Make things easy for the examiner,
Make it clear,
What it is about,
What for,
What the aims are, What did you do?,
What came of it,
What does all that mean,
Conclusions.
Remember the examiner is human, he will have had to put in quite a lot of effort to reach a view of your project and thesis.
Humour him, make his task easy.
Don't demand that he finds a Rosetta Stone to decode your work.
Remember, you wish to do well
     The examiner wishes you to do well
     Your sponsoring body wishes you to do well
     Your University Deptartment wishes you to do well.
If hidden rocks are hit the problem needs to be addressed early and constructively.
Explain early-on why a project to classify black ants ends up as a theory of laminate formation.

Cultural Background

The wheelbarrow test (Physical).

I mentor engineers from many backgrounds and cultures-of-origin.
It is easy to take for granted some of the tacit knowledge, which we have
in our heads.

Imagine the following scenario:

On a building site a set of workmen have dug a trench about 1 metre deep
and 1 metre wide, many metres long.
They have a pile of bricks on one side and need them at a point of use on
the other side of the trench. A wheelbarrow is available.The site is littered
with odd planks and pieces of wood of various sizes.
Based on my childhood experiences working with my grandfather I would
easily select a plank or planks small enough for me to handle, yet strong
enough to allow transit of a barrow of bricks over the trench. Too many
bricks and I could not lift the barrow handles, so the load is finite. A
'found' discarded door would probably serve, so my bricks are soon
moved. Easier than throw-catch brick by brick with a co-worker.
How do I have an intuitive ability to assess the properties of objects on
sight? Tacit knowledge long laid down in early life remains available.
Other persons from different cultures may find that they need a calculator
or extensive trial and error.

Change location to a Research laboratory. A length of wire is needed to
carry 5 amperes for a few minutes.
Based on tacit knowledge and experience a suitable thickness is soon
selected from the reels available – no calculator and no subsequent smoke
– again tacit knowledge from long ago.
It is now clear why hands-on hobby activities contribute so well to the
flexible competence of a young engineer. Get yourself involved in
activities which safely explore mastery of the physical world.

Psycological

My daughter once gave me a black plaster cat with green roll-about eyes
as a birthday gift 'for luck'. On its first day at my office the cleaners
accidentally dropped it and broke its neck
Recalling that 'make your own luck' is positive attitude I fixed the head
back on with glue and seem to have been lucky ever since.

Cross Culture

The cat-donating daughter was keen on horse riding, and on one occasion
we hired two ponies and set off uphill. My pony felt that it was too near
equine lunch time for uphill work and rather than fight my control efforts
merely walked into the close-to-stable duck pond and stood there in 1/2
metre of water and refused to walk out. I value dry feet and was thus
beaten. My daughter, familiar with horse culture merely produced a carrot
and we were soon on dry land again (never did get uphill though!).
Lesson - Respect and understand the cultures you may have to work with.

Appendix

Some Affiliates to the Engineering Council are listed below. A fuller list is available on The Engineering Council Website:
www.org.uk/institutions.aspx

Clearly the Body or Bodies with which you become associated should be relevant to your employment environment or your engagements in Continued Professional Development.

In any case choosing a route towards chartered status is an important move and achievement of chartered status indicates that you have obtained a commendable level of Technological Qualification and Experience.

Once an engineers activities move towards involvement in commercial or contractual situations the possession of Chartered Status can be a valuable, and often essential consideration.
Chartered Status can be quite a large dog snoozing under your desk, so it is well worth finding out what is needed for him to become a friend and servant.
You may never know till too late when the letters CEng would have clinched your job success or promotion.
The Engineering Council UK lists some 30+ bodies on its website.
Among the more valuable to the young engineer are:-

a) The Institution of Engineering and Technology (formerly the Institution of Electrical Engineers).
   Savoy Place, London, WC2R OBL Tel 0207-240-1871.
   www.theiee.org

b) Institution of Civil Engineers
   1-7 Great George St. London SW1P 3AA
   Tel 02072227722
   www.ice.org.uk

c) Institute of Materials, Minerals and Mining
1 Carlton House Terrace, London SW1Y 5DB
Tel 02074517300
www.iom3.org

d) Institute of Measurement and Control
87 Gower Street, London WC1E 6AF
Tel 02073874949
www.instmc.org.uk

e) Institute of Physics
76 Portland place, London W1B 1NT
Tel 02074704800
www.iop.org

f) Institution of Mechanical Engineers
1 Birdcage Walk, London SW1H 9JJ
Tel 02072227899
www.imeche.org.uk

The contact data given were up to date in May 2007 and this book cannot be responsible for later changes.

The message remains – put your degree(s) certificates in a safe place and begin to plan a route to Chartered status which will set you on the road to becoming a truly Effective Engineer.

Bibliography

1.0

How to win friends and influence people
by Dale Carnegie.
Published by Cedar Books and others over many years, inexpensive
paper back versions readily obtainable.
Produced before ISBN numbers current.

2.0

The Mind Map Book
by Tony Buzan
Pub. Plume-Penguin
ISBN 0-452-27322-6.

3.0

The one Minute Manager
by Kenneth Blanchard and Spencer Johnson
Willow Books.
ISBN 0-00-216061-8.

Articles

Training to be the Best by Paul Clapham' IEE Engineering
Management, Feb/ Mar 2005.pp 28-31.
Ibid, How High should you Try?
pp 32-33.

INDEX

Persons Quizzed for data base on Engineer evaluation (Names deleted).

Technical Director – Electric motor Company

CEO of Steel Plant

Commercial Director – Tech. Products Company

Works Manager

Tech. and R and D Manager

Personnel Manager

Consultant

Technical Director

Tech Director of Chem Company

Manager technology

CEO Steel Plant

Manager Gov. Tech Deptartment

Professor, Director of Univ. Eng Deptartment

Research manager (USA)

Welsh Development Agency Director

Prof. Head of Eng School

Prof. Head of Univ. Eng School

Manager Tech. and Research

The Effective Engineer Synopsis

Student Engineers have committed themselves to a stringent course of study aimed at providing a toolkit of technological techniques which will, when supplemented by updates as years pass, support their activities in employment for a lifetime.

Undergraduates are well aware that this means a lengthy commitment to hard work and learning, and they are mentally prepared to undertake it. They will have 'swotted' for A-levels and are ready to apply similar effort to undergraduate engineering. They realise that evening enjoyment and social activity must be tailored so that sufficient effort goes into lectures, assignments and reports.

After 3+ years of effort students expect to get a degree with a respectable grade and to find that this is acceptable to employers.

In the world of sport it is well known that simply running the course is seldom enough and that for stardom a study must be made of what is needed to excel.

To be a sought-after engineer also requires sustained training in relevant techniques.

This little book tries to illustrate the paths leading to excellence.

It looks at the often informal ways in which they are viewed and evaluated by Industry and Academia.

Guidance is given about how to complement academic study with Life Skill related activities. An evaluation is made by questionaire and interview of what employers hope to find and are likely to reward.

This book is intended to aid in the promotion of Engineering in general and thus to raise its status and progress in the world.

The A5 format is produced economically so that it is readily purchased by those it is intended to help.